AI 纯美人物

绘画关键词图鉴 Midjourney 版

AIGC-RY 研究所 著

人民邮电出版社

北 京

图书在版编目（ＣＩＰ）数据

AI纯美人物绘画关键词图鉴 ：Midjourney版 / AIGC-
RY研究所著. -- 北京 ：人民邮电出版社，2023.10
　ISBN 978-7-115-62568-7

　Ⅰ．①A… Ⅱ．①A… Ⅲ．①图像处理软件 Ⅳ.
①TP391.413

中国国家版本馆CIP数据核字(2023)第175717号

内 容 提 要

　　AI是当下无法阻挡的艺术创作趋势。

　　本书首先简要地介绍了一下关键词的使用方法，帮助读者大致了解生成图片的基本原理；正文
实战部分，展示了画面风格、构图方式、人物属性、人物细节、人物服饰、人物动作、场景氛围等
7大主题的AI图片生成效果，并给出了提示词说明，通过图文对应的方式帮助读者了解生成图片的
具体方法，从而生成自己想要的图像。

　　本书适合对AI图像创作感兴趣的读者和有AI图像创作需求的设计师、插画师等阅读。

　◆ 著　　　　　　　　AIGC-RY 研究所
　　　责任编辑　　王　铁
　　　责任印制　　周昇亮

　◆ 人民邮电出版社出版发行　　北京市丰台区成寿寺路 11 号
　　　邮编　100164　　电子邮件　315@ptpress.com.cn
　　　网址　https://www.ptpress.com.cn
　　　雅迪云印（天津）科技有限公司印刷

　◆ 开本：700×1000　1/16
　　　印张：8.5　　　　　　　　2023 年 10 月第 1 版
　　　字数：186 千字　　　　　　2023 年 10 月天津第 1 次印刷

定价：49.80 元
读者服务热线：(010)81055296　印装质量热线：(010)81055316
反盗版热线：(010)81055315
广告经营许可证：京东市监广登字 20170147 号

使用说明

1. 图片效果名称

每张图片效果的中文和英文名称

2. 图片效果

输入关键词会出现的相关图片效果

东京塔　Tokyo Tower ①

②

③ 提示词：Tokyo Tower, vibrant colors, anime style, tall and slender structure, glowing lights, bustling cityscape in the background, fluffy clouds in the sky, a master anime illustrator, ――ar 16:9 ――niji 5

书店　bookstore

④

提示词：bookstore, cozy, bookshelves filled with books, soft lighting, vintage furniture, warm colors, a quiet and peaceful atmosphere, a master anime illustrator, ――ar 16:9 ――niji 5

3. 提示词

达到图片效果所需的全部提示词

4. 关键词

提示词中的关键词用其他颜色标记出来，一目了然

目录
CONTENTS

第1章

画面风格

画面风格是指绘画、摄影、电影或其他视觉媒体中的艺术风格和表现方式，它包括构图、色彩、光影、线条等要素，决定了画面的整体视觉效果。例如，有些作品可能采用写实主义风格，力求展现真实的细节和逼真的形象；而有些作品可能采用卡通风格，突出夸张的特征和生动的色彩。

高动态范围　HDR

提示词：HDR, bright, vibrant colors, surrealistic landscape, dreamy atmosphere, pastel hues，a maste
anime illustrator, ――ar 16:9 ――niji 5

超高清　UHD

提示词：UHD, nature scene, vastness, depth of field, aerial view, high resolution, vibrant colors,
master anime illustrator, ――ar 16:9 ――niji 5

全高清　FHD

提示词：FHD, a dreamy hazy landscape of a river valley surrounded by mountains, in the style of romanticism, with an emphasis on the clouds and misty atmosphere, a master anime illustrator, --ar 16:9 --niji 5

1080P 分辨率　1080P

提示词：1080P, a young girl in a pink dress, standing on a beach, with the sun setting behind her, in the style of romanticism, warm colors, a master anime illustrator, --ar 16:9 --niji 5

2K 分辨率　2K

提示词：2K, a tranquil lake surrounded by mountains, golden ight, shallow depth of field, warm colors, impressionist style, high contrast，a master anime illustrator, --ar 16:9 --niji 5

4K 分辨率　4K

提示词： 4K, a young girl surrounded by a vibrant and colorful forest, with a dreamy look on her face, hazy sunlight filters through the trees and casts shadows on the forest floor,a master anime illustrator, --ar 16:9 --niji 5

8K 分辨率　　8K

提示词：8K, white kitten, in a dreamy atmosphere, with a shallow depth of field, bright and vibrant colors, a master anime illustrator, ——ar 16:9 ——niji 5

高细节　　high detail

提示词：high detail, close-up of a flower in full bloom, pastel colors, light and airy, soft focus, a master anime illustrator, ——ar 16:9 ——niji 5

游戏 CG　　game CG

提示词：game CG, vibrant colors, detailed textures, dynamic poses, intense action, futuristic setting, glowing energy effects, realistic character models, cinematic composition, epic battle scene, a master anime illustrator, ––ar 16:9 ––niji 5

官方艺术　　official art

提示词：official art, vibrant colors, dynamic poses, detailed backgrounds, character interactions, expressive facial expressions, a master anime illustrator, ––ar 16:9 ––niji 5

动画截图　anime screencap

提示词：anime screencap, colorful and vibrant, dynamic action scene, intense expressions, dramatic lighting, intricate background details, stylized character designs, a master anime illustrator, ——ar 16:9 ——niji 5

漫画　comic

提示词：comic, colorful, dynamic poses, exaggerated expressions, action lines, vibrant backgrounds, comic book style, inked outlines, bold colors, a master anime illustrator, ——ar 16:9 ——niji 5

立绘样式　tachi-e

提示词：tachi-e, traditional Japanese standing figure painting, intricate details, vibrant colors, dynamic poses, flowing robes, delicate brushwork, realistic facial expressions, traditional Japanese setting, a master anime illustrator, --ar 16:9 --niji 5

像素风　pixel art

提示词：pixel art, vibrant colors, retro aesthetic, 8-bit style, detailed and intricate designs, nostalgic feel, blocky pixels, limited color palette, dynamic composition, iconic characters, nostalgic video game vibes, --ar 16:9 --niji 5

海报　poster

提示词：poster, colorful, abstract shapes, eye-catching design, vibrant colors, dynamic composition, modern style, a master anime illustrator, --ar 16:9 --niji 5

马克笔风格　marker style

提示词：marker style, vibrant colors, bold stroke of color, expressive lines, abstract shapes, dynamic composition, energetic atmosphere, graffiti-inspired, street art, mixed media, a master anime illustrator, --ar 16:9 --niji 5

日本浮世绘　Japanese Ukiyo-e

提示词：Japanese Ukiyo-e, muted colors, minimalistic design,Mount Fuji in the distance, everyday life scenes, subtle brushstrokes, tranquil and contemplative mood, by Kitagawa Utamaro, --ar 16:9 --niji 5

日本漫画风格　Japanese manga

提示词：Japanese manga, colorful and vibrant illustrations, dynamic action scenes, expressive characters with large eyes and exaggerated features, intricate backgrounds filled with details, a master anime illustrator, --ar 16:9 --niji 5

日本动画片　anime

提示词：anime, cute chibi characters, pastel colors, soft and dreamy backgrounds, adorable expressions, playful interactions, fluffy animals companions, magical transformations, heartwarming stories, a master anime illustrator, ——ar 16:9 ——niji 5

日本海报风格　poster of Japanese graphic design

提示词：poster of Japanese graphic design, pastel colors, delicate illustrations, dreamy atmosphere, soft gradients, flowing lines, intricate details, , a master anime illustrator, ——ar 16:9 ——niji 5

宫崎骏风格　Miyazaki Hayao style

提示词：**Miyazaki Hayao style, whimsical and enchanting, vibrant colors, fantastical creatures, lush landscapes, magical adventures, gentle and heartwarming storytelling, hand-drawn animation, intricate details, nostalgic atmosphere, --ar 16:9 --niji 5**

新海诚风格　Makoto Shinkai style

提示词：Makoto Shinkai style, girl, beautiful, flowing hair, radiant smile, elegant dress, holding a bouquet of flowers, in a sunlit garden, surrounded by colorful butterflies, soft and dreamy atmosphere, --ar 16:9 --niji 5

副岛成记风格　Soejima Shigenori style

提示词：**Soejima Shigenori style, men, handsome, muscular, confident, wearing a tailored suit, standing against a city skyline, evening light, cinematic style, ––ar 16:9 ––niji 5**

山田章博风格　Yamada Akihiro style

提示词：**girl, innocent and pure look, Yamada Akihiro style, soft pastel colors, whimsical elements, fantasy setting with butterflies, wearing a flowing dress and flower crown, surrounded by nature and magical creatures, ––ar 16:9 ––niji 5**

吉卜力风格　　Ghibli Studio style

提示词：Ghibli Studio style, dense forest, towering trees, mysterious atmosphere, birds chirping
gentle breeze rustling the leaves, hidden creatures lurking in the shadows, ethereal an
enchanting, water droplets hanging from the branches, vibrant shades of green, --ar 16:9 --niji 5

二次元　　ACGN

提示词：ACGN, colorful characters, dynamic poses, vibrant backgrounds, intricate details, anim
style, --ar 16:9 --niji 5

卡通　cartoon

提示词：**cartoon, colorful, whimsical characters, exaggerated features, vibrant backgrounds, playful and lively scenes, comic book style, bold outlines, dynamic poses, expressive facial expressions, --ar 16:9 --niji 5**

星际战甲风格　warframe style

提示词：warframe style, advanced exoskeleton suits, customizable appearance, intricate designs, vibrant color schemes, aerial combat, alien landscapes, massive boss battles, cinematic cutscenes, --ar 16:9 --niji 5

魂系游戏　From Software

提示词：From Software, video game, dark and atmospheric, challenging gameplay, intricate leve design, intense boss battles, mysterious storyline, gothic architecture, medieval setting, --ar 16: --niji 5

微缩模型电影风格　miniature movie style

提示词：miniature movie style, intricate set design, tiny detailed props, dramatic lighting, cinemati camera angles, shallow depth of field, vintage color grading, nostalgic atmosphere, storytellin composition, --ar 16:9 --niji 5

第 2 章

构图方式

构图方式是指摄影或绘画作品中所采用的观察角度，或者是场景和主题的呈现方式。它对整体画面的表现力和视觉效果有着重要影响，也会给读者带来不同的观感和情感体验。我们可以根据主题和表现意图选择合适的构图方式。

饱和构图　saturated composition

提示词：saturated composition, vibrant colors, contrasting elements, dynamic movement, bold shapes energetic atmosphere, expressive lines, a master anime illustrator, ––ar 1:2 ––niji 5

视点构图　point of view composition

提示词：point of view composition, an abandoned city, dark and gloomy atmosphere, with a hin of mystery, streetlights in the distance, neon lights in the foreground, a sense of despair, a maste anime illustrator, ––ar 1:2 ––niji 5

剪影构图　cut out composition

提示词：cut out composition, a person standing in silhouette against the warm-lit sky at sunset, shallow depth of field, a master anime illustrator, --ar 1:2 --niji 5

重复构图　repetition composition

提示词：repetition composition, abstract geometric shapes, minimalism style, with blue and yellow colors, very detailed, a master anime illustrator, --ar 1:2 --niji 5

焦点构图　focal point composition

提示词：focal point composition, a landscape with a mountain peak and a lake, light and shadows, pastel colors, high contrast, a master anime illustrator, --ar 1:2 --niji 5

对比构图　contrast composition

提示词：contrast composition, two elements in a frame, one in the foreground and the other in the background, with a strong contrast between them, minimalistic style, --ar 1:2 --niji 5

重叠构图　overlapping composition

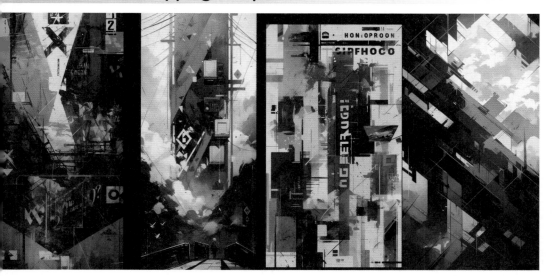

提示词：overlapping composition, monochromatic tones, geometric patterns, strong shadows and highlights, high contrast, industrial setting, grungy texture, vintage style, in a rectangular format, master anime illustrator, --ar 1:2 --niji 5

孤立构图　isolation composition

提示词：isolation composition, stark contrast, monochromatic color scheme, sharp lines and angles, negative space, high-key lighting, --ar 1:2 --niji 5

径向构图　　radial composition

提示词：radial composition, floral patterns, pastel colors, organic shapes, soft and delicate edge: romantic atmosphere, vintage texture, natural lighting, a master anime illustrator, --ar 1:2 --niji

黄金分割构图　　Golden Ratio composition

提示词：Golden Ratio composition, serene mountain landscape, misty atmosphere, pastel color palett soft and delicate brushstrokes, a master anime illustrator, --ar 1:2 --niji 5

对角线构图　diagonal composition

提示词：diagonal composition, vintage feel, classic car, chrome accents, sleek lines, dramatic shadows and highlights, film noir atmosphere, ––ar 1:2 ––niji 5

对称构图　symmetrical composition

提示词：symmetrical composition, minimalist design, monochromatic color scheme, sharp lines and angles, geometric shapes, high contrast lighting, futuristic atmosphere, ––ar 1:2 ––niji 5

顶视图　top view

提示词：**top view**, bird's eye perspective, aerial shot, overlooking, panoramic view, vast landscape, cityscape, buildings, streets, cars, people, bustling activity, vibrant colors, dynamic composition, urban life, energetic atmosphere, −−ar 1:2 −−niji 5

前视图　front view

提示词：front view, girl, standing posture, confident expression, flowing hair, casual outfit, vibrant colors, natural lighting, outdoor setting, composition focused on the subject, energetic atmosphere, −−a 1:2 −−niji 5

侧面视角　from side

提示词: from side, anime characters, intricate hairstyles and accessories, soft lighting, dreamy atmosphere, pastel colors, floral patterns, elegant poses and expressions, watercolor style, by Makoto Shinkai, --ar 1:2 --niji 5

后视图　back view

提示词: back view, a girl standing in a natural landscape, a master anime illustrator, --ar 1:2 --niji 5

从上方看　from above

提示词：from above, capturing a bird's eye view, creating a sense of dominance and power, emphasizin
the subject's presence and importance in the scene, a master anime illustrator, −−ar 1:2 −−niji 5

从下方看　from below

提示词：from below, towering skyscrapers, reaching towards the sky, glass and steel facade, reflectin
the clouds above, bustling city life below, birds soaring overhead, −−ar 1:2 −−niji 5

两点透视　two-point perspective

示词：two-point perspective, geometric shapes, sharp edges, clean lines, minimalistic color palette, contrasting shadows, futuristic atmosphere, high contrast, --ar 1:2 --niji 5

三点透视　three-point perspective

示词：three-point perspective, streetscape,clean lines, a master anime illustrator, --ar 1:2 --niji 5

头部特写　head shot

提示词：head shot, cute and innocent anime characters, pastel color palette, soft and dreamy atmospher big expressive eyes, blush on the cheeks, floating bubbles or sparkles around the character, in watercolor style, --ar 1:2 --niji 5

脸部特写　face shot

提示词：face shot, anime girl, soft pastel color palette, gentle expression, floral hair accessory, dream atmosphere, watercolor texture, Ghibli Studio style,--ar 1:2 --niji 5

胸部以上　　chest shot

提示词：chest shot, cute and innocent anime characters, pastel color palette, soft and dreamy atmosphere, master anime illustrator, ––ar 1:2 ––niji 5

膝盖以上　　knee shot

提示词：knee shot, fitted and sleek cocktail dress, standing on a rooftop overlooking the city skyline at night, confident and sultry expression, ––ar 1:2 ––niji 5

半身像　busts

提示词：**busts, lifelike expressions and details, dramatic lighting, classical Greek style, intricate hair and clothing details, ––ar 1:2 ––niji 5**

全身　full body

提示词：full body, ethereal goddess, flowing white dress, billowing chiffon, golden halo, surrounded by mist and flowers, soft pastel colors, dreamlike atmosphere, ––ar 1:2 ––niji 5

细节镜头　detail shot

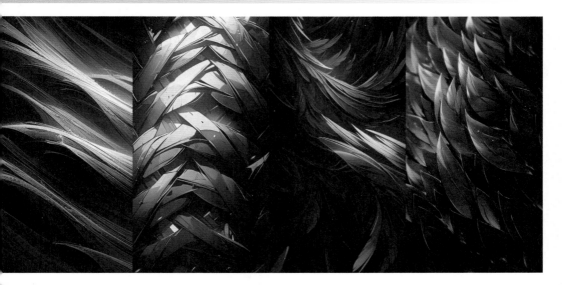

提示词：detail shot, extreme close-up, intricate textures and patterns, shallow depth of field, dramatic lighting, contrasting colors, minimalist composition, abstract and surreal atmosphere, --ar 1:2 --niji 5

背景虚化　bokeh

提示词：bokeh, soft and ethereal, pastel colors, flowers in the foreground, blurred background, anime character, whimsical and playful expression, holding a string of balloons, in a magical forest setting, --ar 1:2 --niji 5

中景　medium shot

提示词：medium shot，a woman, wearing a navy blue suit and a red tie, standing in front of a modern skyscraper, the sun shining brightly in the background, a master anime illustrator, ––ar 1:2 ––niji 5

远景　extra long shot

提示词：extra long shot, silhouettes against a fiery sunset, hazy and dreamy atmosphere, palm trees swaying in the wind, warm colors of orange, pink and yellow, composition with negative space reminiscent of a postcard from a tropical paradise, ––ar 1:2 ––niji 5

第 3 章

人物属性

人物属性既包括角色的性别、年龄、身高、体形等生理属性，也包括角色的性格、价值观、能力等心理属性。人物属性可以定义角色的身份、个性和背景，使他们在故事中具有独特性和可识别性。

单人　solo

提示词：solo, anime style, vibrant colors, dynamic pose, flowing hair, expressive eyes, soft lighting magical atmosphere, a master anime illustrator, --ar 1:2 --niji 5

女人　female

提示词：female, wise and knowledgeable, reading glasses, thoughtful expression, cozy study room with bookshelves filled with books, warm and inviting ambiance, a master anime illustrator, --a 1:2 --niji 5

男人　male

提示词：male, intense gaze, short hair, wearing a leather jacket and jeans, a master anime illustrator, ––ar 1:2 ––niji 5

婴儿　baby

提示词：baby, adorable, chubby cheeks, tiny fingers and toes, dressed in a cute onesie, surrounded by colorful toys, giggling and playing, soft and gentle lighting, captured in a candid moment, a master anime illustrator, ––ar 1:2 ––niji 5

儿童　child

提示词：child, dressed in a superhero costume with a cape flowing behind them, a master anime illustrator, ––ar 1:2 ––niji 5

青年男孩　teenage boy

提示词：teenage boy, anime style, messy hair, glasses, casual clothes with a hoodie, a master anime illustrator, ––ar 1:2 ––niji 5

成熟女性　mature female

提示词：mature female, graceful and poised, wearing a flowing dress, with a serene expression and gentle smile, sitting in a sunlit garden surrounded by colorful flowers, a master anime illustrator, ––ar 1:2 ––niji 5

老年人　elderly individual

提示词：elderly individual, anime style, wrinkled skin, gray hair, glasses, walking canes, wise and kind expressions, nostalgic and heartwarming atmosphere, a master anime illustrator, ––ar 1:2 ––niji 5

2 个女孩　2 girls

提示词：**2 girls** in an anime style, monochromatic color scheme with pops of vibrant colors, intricate and detailed outfits, contrasting personalities, in a gothic lolita style dress with dark makeup and accessories, dramatic lighting casting shadows on their faces, a master anime illustrator, ––ar 1:2 ––niji 5

2 个男孩　2 boys

提示词：**2 boys**, anime style, best friends, with a strong bond and deep connection, one with a protective and caring expression, the other with a playful and cheerful smile, matching outfits that symbolize their friendship, a master anime illustrator, ––ar 1:2 ––niji 5

母子　mother and children

提示词：mother and children, anime style, gentle expressions, flowing hair, big sparkling eyes, cute outfits, surrounded by cherry blossom trees, a warm and sunny day, in a peaceful park, a master anime illustrator, ––ar 1:2 ––niji 5

Q版　chibi

提示词：chibi, adorable and mischievous, chubby cheeks, messy hair, wearing oversized glasses, a master anime illustrator, ––ar 1:2 ––niji 5

萝莉　loli

提示词：loli, cute, kawaii, pastel colors, big eyes, pigtails, frilly dress, soft and dreamy atmosphere, a master anime illustrator, ——ar 1:2 ——niji 5

正太　shota

提示词：shota, mischievous, messy hair, cheeky smile, holding a slingshot, surrounded by lush greenery, sunlight filtering through the leaves, warm and vibrant colors, a master anime illustrator, ——ar 1:2 ——niji 5

美少女　bishoujo

提示词：bishoujo, elegant, flowing dress, long flowing hair, delicate features, radiant smile, soft and natural lighting, a master anime illustrator, ––ar 1:2 ––niji 5

人外娘　monster girl

提示词：monster girl, fierce and powerful, sharp claws and teeth, glowing eyes, colorful scales or fur, horns or antlers, flowing mane or tail, in a mystical forest, surrounded by magical creatures, ethereal and dreamlike atmosphere, painted with vibrant colors and intricate details, a master anime illustrator, ––ar 1:2 ––niji 5

兔娘　rabbit girl

提示词：rabbit girl, adorable, fluffy white fur, long ears, cute button nose, hopping playfully in a grassy meadow, surrounded by colorful flowers, soft sunlight filtering through the trees, a master anime illustrator, --ar 1:2 --niji 5

狐娘　fox girl

提示词：fox girl, standing posture, mischievous and playful expression, short orange fur, perky ears and a fluffy tail, wearing a cute and colorful outfit with matching accessories, standing in a field of wildflowers, butterflies fluttering around her, vibrant and cheerful color palette, whimsical atmosphere, a master anime illustrator, --ar 1:2 --niji 5

猫娘　　cat girl

提示词：cat girl, elegant, graceful, wearing a flowing dress with a flower print, long flowing hair with cat ears,standing in a garden, a master anime illustrator, ‒‒ar 1:2 ‒‒niji 5

犬娘　　dog girl

提示词：dog girl, playful, bow on its head, happy and carefree expression, soft and fluffy fur, ‒‒ar 1:2 ‒‒niji 5

美人鱼　mermaid

提示词：mermaid, ethereal beauty, flowing hair, shimmering scales, underwater scene, soft pastel colors, dreamy atmosphere, water droplets glistening on skin, graceful swimming motion, a master anime illustrator, --ar 1:2 --niji 5

机娘　mecha musume

提示词：mecha musume, anime style, cute and chibi version, pastel colors, oversized weapon prop, playful and cheerful expression, floating in mid-air with wings, fluffy clouds in the background, a master anime illustrator, --ar 1:2 --niji 5

小精灵　fairy

提示词：fairy, mischievous and playful, short pixie haircut with vibrant colored highlights, wearing a whimsical outfit with mismatched patterns and textures, holding a jar of glowing fireflies, surrounded by magical floating orbs, in a enchanted garden at twilight, a master anime illustrator, --ar 1:2 --niji 5

妖精　elf

提示词：elf, mischievous trickster, short stature, mischievous grin, colorful clothing with mismatched patterns, playing pranks on unsuspecting creatures in a whimsical forest, surrounded by magical creatures like fairies and talking animals, playful and lighthearted atmosphere, a master anime illustrator, --ar 1:2 --niji 5

魔法少女　magical girl

提示词：magical girl, standing posture, sparkles in the air, colorful and vibrant outfit, long flowing hair, confident expression, surrounded by a halo of light, in a mystical forest, with butterflies and flowers around her, --ar 1:2 --niji 5

矮人　dwarf

提示词：dwarf, small stature, long beard, pointy hat, carrying a pickaxe, deep wrinkles on face, wise and knowledgeable expression, living in a cozy underground cave, surrounded by precious gemstones and minerals, soft warm lighting, fantasy setting, a master anime illustrator, --ar 1:2 --niji 5

第 4 章

人物细节

人物细节是指角色形象的具体特征和细节描述。它包括了角色的外貌特征，如面部特征、发型、身材等，以及身上的标志性特征，如疤痕、文身、特殊的服饰等。人物细节的描写可以帮助读者更好地想象和理解角色，以及与故事情节相关的细节。

明亮的眼睛　　light eyes

提示词: girl with light eyes, anime style, colorful hair, vibrant and expressive eyes, cute and innocent smile, with soft pastel colors, a master anime illustrator, ––ar 2:2 ––niji 5

渐变眼睛　　gradient eyes

提示词: gradient eyes, close-up of face, anime style, large sparkling eyes, vibrant colors, soft and smooth skin, expressive eyebrows, delicate eyelashes, rosy cheeks, a master anime illustrator, ––ar 2:2 ––niji 5

动画眼　anime style eyes

提示词：anime style eyes, close-up of face, soft and gentle, round and innocent pupils, delicate and wispy lashes, subtle pink highlights, shy and sweet expression, surrounded by a serene and peaceful landscape with cherry blossom trees in the background, in a romantic and whimsical style, --ar 2:2 --niji 5

实心椭圆眼睛　solid oval eyes

提示词：solid oval eyes, close-up of face, anime style, big round glasses, colorful hair, vibrant and expressive eyes, soft and smooth skin, cute and innocent expression, blush on cheeks, a master anime illustrator, --ar 2:2 --niji 5

爱心形瞳孔　heart-shaped pupils

提示词：**heart-shaped pupils, close-up of face, anime style, big expressive eyes, rosy cheeks, long lashes, shiny hair, vibrant colors, cute and innocent expression, a master anime illustrator, --ar 2:2 --niji 5**

竖的瞳孔　slit pupils

提示词：**slit pupils, mysterious and alluring, long flowing hair, wearing a black lace dress, composition centered on the girl's face, a master anime illustrator, --ar 2:2 --niji 5**

闭眼 　eyes closed

提示词: eyes closed, peaceful expression, serene atmosphere, soft lighting, gentle breeze, flowing hair, tranquil setting, a master anime illustrator, ––ar 2:2 ––niji 5

闪光的眼睛 　sparkling eyes

提示词: anime character, sparkling eyes, close–up of face, intense gaze, vibrant colored hair, flawless skin, expressive eyes, detailed eyelashes and eyebrows, dramatic lighting, glowing eyes, mysterious and captivating expression, in a fantasy setting, digital painting style, by Studio Ghibli, ––ar 2:2 ––niji 5

紫色瞳孔　purple pupils

提示词：**girl with purple pupils, mysterious, long flowing hair, a master anime illustrator, ––ar 2:2 ––niji 5**

蓝色瞳孔　blue pupils

提示词：**girl with blue pupils, mysterious and enchanting, long flowing hair, a master anime illustrator, ––ar 2:2 ––niji 5**

没鼻子　no nose

提示词：**no nose, close-up shot, anime style, bold and vibrant colors, exaggerated features, wide eyes and bright lips, dynamic and energetic expression, a master anime illustrator, --ar 2:2 --niji 5**

点状鼻　dot nose

提示词：**dot nose, close-up of face, anime style, large expressive eyes, rosy cheeks, vibrant hair color, intricate hair accessories, soft pastel colors, a master anime illustrator, --ar 2:2 --niji 5**

努嘴　pout

提示词：**pout, pouty expression, close-up of face, vibrant and bold colors, anime style, large eyes with thick lashes, glossy lips, tousled hair, unique and intricate earrings, by Hayao Miyazaki, ––ar 2:2 ––niji 5**

圆齿　round teeth

提示词：**round teeth girl, anime style, big sparkling eyes, rosy cheeks, a master anime illustrator, ––ar 2:2 ––niji 5**

锋利的牙齿　sharp teeth

提示词：cute girl with **sharp teeth**, mischievous smile, long flowing hair, a master anime illustrator, ––ar 2:2 ––niji 5

虎牙　fangs

提示词：cute girl with **fangs**, mischievous smile, long flowing hair, a master anime illustrator, ––ar 2:2 ––niji 5

狐狸耳朵　　fox ears

提示词：**fox ears** girl with sleek silver hair and piercing blue eyes, exuding an aura of mystery and elegance, a master anime illustrator, ––ar 2:2 ––niji 5

猫耳朵　　cat ears

提示词：cat ears girl with sleek hair, mysterious anime style, a master anime illustrator, ––ar 2:2 ––niji 5

兔耳　bunny ears

提示词：**bunny ears** girl, anime style, lovely, pastel colors, oversized bow, big round eyes, a master anime illustrator, ––ar 2:2 ––niji 5

蝙蝠耳朵　bat ears

提示词：bat ears girl, anime style, vibrant colors, flowing hair, playful expression, a master anime illustrator, ––ar 2:2 ––niji 5

机器人耳朵　robot ears

提示词：boy with robot ears, anim style, futuristic clothing, vibran colors, expressive eyes, dynami pose, a master anime illustrato --ar 2:2 --niji 5

长尖耳朵　long pointy ears

提示词：elf girl with long pointy ears, pastel hair colors, big expressive eyes, delicate features, intricate jewelry, flowers in hair, soft and dreamy atmosphere, watercolor style, by Makoto Shinkai, --ar 2:2 --niji 5

动画式脸红　anime style blush

提示词：anime style blush girl with rosy cheeks and a shy smile, a master anime illustrator, ar 2:2 niji 5

咧嘴笑　grin

提示词：grin, closeup portrait, anime style, vibrant and lively expression, bright and sparkling eyes, infectious smile, flushed cheeks, a master anime illustrator, ar 2:2 niji 5

魅惑的微笑　seductive smile

提示词: seductive smile, close-up of face, anime style, mysterious and alluring expression, piercing eyes, dark and smoky makeup, sleek and shiny hair, a master anime illustrator, --ar 2:2 --niji

泪如雨下　streaming tears

提示词: streaming tears, close-up of face, anime style, big teardrops, sparkling and shimmering, emotional expression, vibrant colors, soft and glowing skin, flowing hair, tear-streaked cheeks, watery eyes, delicate and detailed linework, a master anime illustrator, --ar 2:2 --niji 5

烦恼　　annoyed

提示词：annoyed girl, furrowed brows, messy hair, pastel color palette, soft and dreamy atmosphere, a master anime illustrator, ––ar 1:1 ––niji 5

开心的眼泪　　happy tears

提示词：happy tears, boy, joyful, tears of joy streaming down his face, wide smile, hands raised in celebration, a master anime illustrator, ––ar 2:2 ––niji 5

害羞的　shy

提示词：shy girl, close-up of face, manga style, small blushing cheeks, innocent expression, large round eyes with starry sparkle, long eyelashes, soft pastel colors, flowing hair with ribbons and bows, a master anime illustrator, --ar 2:2 --niji 5

捂脸　facepalm

提示词: facepalm, close-up of face, anime style, dramatic lighting, shadow cast over half the face, a master anime illustrator, --ar 2:2 --niji 5

流汗　sweat

提示词: sweat face, up-close view, manga style, beads of sweat, dewy skin, determined expression, a master anime illustrator, --ar 2:2 --niji 5

害怕的　scared

提示词: scared, close-up of face, anime style, wide eyes with dilated pupils, trembling lips and chin, sweat dripping from forehead, tears welling up in eyes, pale complexion, messy hair strands sticking out, a master anime illustrator, --ar 2:2 --niji 5

傲娇　tsundere

提示词：**tsundere, close-up of face, anime style, blushing cheeks, sharp eyes, pouting lips, flushed expression, colorful hair, soft lighting, a master anime illustrator, --ar 2:2 --niji 5**

直发　straight hair

提示词：straight hair, **close-up of face, anime style, soft pastel colors, gentle expression, serene atmosphere, floral background, a master anime illustrator, --ar 2:2 --niji 5**

卷发　curly hair

提示词：curly hair, close-up of face, anime style, big expressive eyes, vibrant colors, soft and smooth skin, rosy cheeks, glossy lips, flowing strands of hair, sparkling highlights, delicate eyelashes, a master anime illustrator, --ar 2:2 --niji 5

双马尾　twintails

提示词：girl with twintails, vibrant colors, long and curly, a master anime illustrator, --ar 2:2 --niji 5

姬发式　hime cut

提示词: hime cut, anime style, big sparkling eyes, rosy cheeks, flawless skin, pastel hair color, delicate flower crown, soft and dreamy lighting, ethereal atmosphere, a master anime illustrator, --ar 2:2 --niji 5

马尾辫　ponytail

提示词: ponytail, anime style, sleek and polished hair, natural and soft makeup, pastel colors, gentle and diffused lighting, dreamy and ethereal atmosphere, a master anime illustrator, --ar 2:2 --niji 5

长辫子　long braid

提示词：**long braid** girl, flowing hair, vibrant colors, Bohemian style, flower crown, flowing dress, golden hour lighting, dreamy and ethereal atmosphere, a master anime illustrator, ––ar 2:2 ––niji 5

法式冠编发　crown braid

提示词：**crown braid**, close–up of face, anime style, sleek and polished braiding, pearls and gemstones woven into the braid, flawless and porcelain–like skin, intense and piercing eyes, sculpted cheekbones with a hint of contour, bold and vibrant lipstick, straight and sleek bangs, a master anime illustrator, ––ar 2:2 ––niji 5

丸子头　bun

提示词：bun girl, playful and energetic, colorful hair accessories, a master anime illustrator, --ar 2:2 --niji 5

眼睛之间的头发　hair between eyes

提示词：hair between eyes, in an anime style, soft and muted colors, mysterious expression, a master anime illustrator, --ar 2:2 --niji 5

头发覆盖一只眼　hair over one eye

提示词：hair over one eye, rebellious, vibrant color, messy and tousled, wind blowing through the hair, a master anime illustrator, --ar 2:2 --niji 5

短发　short hair

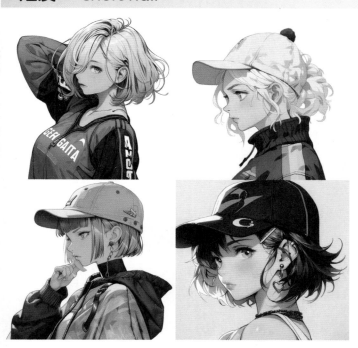

提示词：short hair girl, sporty, a master anime illustrator, --ar 2:2 --niji 5

黑色头发　black hair

提示词：**girl with** black hair, **anime style, big eyes, long flowing hair, colorful hair accessories, cute expression, soft and dreamy atmosphere, a master anime illustrator, --ar 2:2 --niji 5**

棕色头发　brown hair

提示词：brown haired **girl, curly hair, wearing glasses, a master anime illustrator, --ar 2:2 --niji 5**

绿色头发　green hair

提示词: **girl with green hair, vibrant and neon, long and flowing, styled in intricate braids, surrounded by lush green foliage, standing in a mystical forest, soft sunlight filtering through the trees, a master anime illustrator, --ar 2:2 --niji 5**

蓝色头发　blue hair

提示词: **blue hair girl, ethereal blue hair cascading down her back, wearing a flowing white dress, a master anime illustrator, --ar 2:2 --niji 5**

银发　silver hair

提示词: **silver haired** protagonist, close-up shot of the face, anime style, gentle smile, rosy cheeks, sparkling eyes, flowing strands of hair, soft and warm lighting, a master anime illustrator, --ar 2:2 --niji 5

金发　blonde hair

提示词: girl with fiery **blonde hair** cascading down her back, focus on close-up of face, in an anime style, striking eyes, fair complexion, plump pink lips, delicately arched eyebrows, soft blush on the cheeks, mischievous grin with a hint of playfulness, a master anime illustrator, --ar 2:2 --niji 5

粉色头发　pink hair

提示词：**pink hair, close-up of face, anime style, soft and pastel colors, dreamy and ethereal atmosphere, gentle and innocent expression, sparkling eyes, a master anime illustrator, --ar 2:2 --niji 5--ar 2:3 --niji 5**

红色头发　red hair

提示词：**girl with fiery red hair cascading down her back, focus on close-up of face, in an anime style, striking eyes, fair complexion, plump pink lips, delicately arched eyebrows, soft blush on the cheeks, mischievous grin with a hint of playfulness, a master anime illustrator, --ar 2:2 --niji 5**

两色头发　two-tone hair

提示词: **two-tone hair, focus on close-up of face, in an anime style, striking green eyes, fair complexion, plump pink lips, delicately arched eyebrows, soft blush on the cheeks, mischievous grin with a hint of playfulness, a master anime illustrator, --ar 2:2 --niji 5**

彩虹色头发　rainbow colored hair

提示词: **rainbow colored hair, vibrant colors, flowing and wavy, cascading down the shoulders, shimmering under the sunlight, magical and ethereal, like a unicorn's mane, in a whimsical and dreamy setting, surrounded by colorful flowers and butterflies, with a soft and gentle breeze blowing through, a master anime illustrator, --ar 2:2 --niji 5**

苍白皮肤　pale skin

提示词：pale skin, porcelain-like complexion, soft and smooth texture, subtle blush on the cheeks, delicate and refined features, ethereal and otherworldly beauty, natural light illuminating the face, a master anime illustrator, ——ar 2:2 ——niji 5

白皙皮肤　fair skin

提示词：fair skin, flawless complexion, soft and smooth, youthful appearance, gentle and delicate, natural beauty, soft sunlight casting a warm glow, a master anime illustrator, ——ar 2:2 ——niji 5

棕色皮肤　brown skin

提示词：brown skin, glowing under the warm sunlight, smooth and flawless texture, natural and radiant beauty, a master anime illustrator, --ar 2:2 --niji 5

黑色皮肤　black skin

提示词：black skin, rich and deep complexion, in a majestic pose that commands attention, a master anime illustrator, --ar 2:2 --niji 5

第5章

人物服饰

人物服饰是指角色所穿着的衣物和佩戴的配饰，可以很好地展示角色的个性、社会地位、文化背景和时代背景等方面的信息。我们可以通过服饰的材质、配色等方面来传达出角色的职业、气质、情感状态等信息。

衬衫　　shirt

提示词：shirt, boy, casual, short sleeves, vibrant colors, relaxed fit, sunny outdoor scene, a master anime illustrator, --ar 1:2 --niji 5

水手服　　sailor suit

提示词：sailor suit, vibrant colors, anime style, oversized collar, puffy sleeves, short length, sailor collar with a bow, pleated skirt, playful expression, dynamic pose, a master anime illustrator, --ar 1:2 --niji 5

背心式紧身衣　camisole

提示词：**camisole, anime style, vibrant colors, bold patterns, crop top design, high-waisted skirt, playful pose, a master anime illustrator, --ar 1:2 --niji 5**

雨衣　raincoat

提示词：**raincoat, anime-inspired design, oversized hood, waterproof material, translucent and shiny, bold patterns and prints, a master anime illustrator, --ar 1:2 --niji 5**

套头毛衣　pullover sweaters

提示词：**pullover sweaters, anime style, earthy tones, loose and cozy fit, nature-inspired patterns, characters in peaceful and serene settings, a master anime illustrator, --ar 1:2 --niji 5**

露肩连衣裙　off-shoulder dress

提示词：**off-shoulder dress, vibrant colors, flowing fabric, floral print, cinched waist, flared skirt, bare shoulders, in confident pose, under natural sunlight, outdoor setting, bohemian atmosphere, a master anime illustrator, --ar 1:2 --niji 5**

婚纱　wedding dress

提示词：**wedding dress, flowing and ethereal, lace details, intricate embroidery, sweetheart neckline, voluminous skirt, fairy tale setting, blooming garden, golden sunlight streaming through trees, dreamy atmosphere, a master anime illustrator, ––ar 1:2 ––niji 5**

短裤　shorts

提示词：**shorts, anime style, vibrant colors, fitted silhouette, playful patterns, cute accessories, oversized shirt, sunlit park, by a pond with Koi Fish swimming, a master anime illustrator, ––ar 1:2 ––niji 5**

南瓜裤　pumpkin pants

提示词：pumpkin pants, vibrant and playful colors, oversized and baggy fit, cute and whimsical patterns, paired with a cropped top or a fitted blouse, a master anime illustrator, ––ar 1:2 ––niji 5

连帽斗篷　hooded cloak

提示词：hooded cloak, girl, elegant, flowing fabric, vibrant colors, enchanted garden, sunlight streaming through trees, a master anime illustrator, ––ar 1:2 ––niji 5

冬季大衣　winter coat

提示词：**winter coat, cozy, faux fur trim, neutral color palette, oversized hood, chunky knit scarf, leather gloves, snow-covered landscape, soft falling snowflakes, warm and inviting atmosphere, classic style, a master anime illustrator, --ar 1:2 --niji 5**

西装　suit

提示词：**suit, tailored, slim fit, classic black, crisp white shirt, polished shoes, elegant, professional, confident, a master anime illustrator, --ar 1:2 --niji 5**

和服　kimono

提示词：kimono, long flowing hair, delicate floral patterns,surrounded by cherry blossom trees, soft sunligh filtering through the leaves, peaceful and serene atmosphere, traditional Japanese architecture in the background, a master anime illustrator, ––ar 1:2 ––niji 5

厨师工装　chef uniform

提示词：chef uniform, anime style, tall chef hat, crisp white double–breasted jacket, shiny black shoes, holding a wooden spoon and a frying pan, surrounded by floating ingredients like vegetables and spices, vibrant colors, a master anime illustrator, ––ar 1:2 ––niji 5

实验服　labcoat

提示词：labcoat, anime style, vibrant colors, flowing hair, intense eyes, dynamic pose, futuristic laboratory setting, glowing test tubes, scientific equipment, digital interface, a master anime Illustrator, ──ar 1:2 ──niji 5

运动服　sportswear

提示词：sportswear, vibrant colors, sleek design, breathable fabric, form-fitting, a master anime Illustrator, ──ar 1:2 ──niji 5

哥特风格　gothic

提示词: gothic anime style, mysterious and dark atmosphere, intricate and detailed costumes, dramatic poses and expressions, flowing hair with ornate accessories, ethereal and haunting mood, a master anime illustrator, ––ar 1:2 ––niji 5

风衣　trench coat

提示词: trench coat, classic, belted, neutral color, tailored fit, double-breasted, wide lapels, buttoned cuffs, sophisticated and elegant, a master anime illustrator, ––ar 1:2 ––niji 5

睡衣　sleepwear

提示词：sleepwear, cozy and soft, pastel colors, floral patterns, satin fabric, loose fit, comfortable pajamas, in a bedroom with warm lighting and plush carpet, a master anime illustrator, ––ar 1:2 ––niji 5

樱丘女子高等学校校服　Sakuragaoka high school uniform

提示词：Sakuragaoka high school uniform, anime style, plaid skirt, fitted blazer, necktie, dark color palette, a master anime illustrator, ––ar 1:2 ––niji 5

夹克　jacket

提示词：**jacket, stylish, leather, fitted, asymmetrical zipper, metallic accents, biker-inspired, edgy vibe, black and silver color scheme, high collar, quilted details, a master anime illustrator, --ar 1:2 --niji 5**

女仆装　maid

提示词：**maid, elegant, black and white uniform, confident posture, serving tray with a cup of tea, vintage-inspired setting, soft lighting, a master anime illustrator, --ar 1:2 --niji 5**

泳装　swimsuit

提示词：**swimsuit, vibrant colors, tropical print, high-cut leg, strappy details, beach scene, crystal clear water, a master anime illustrator, --ar 1:2 --niji 5**

沙滩裤　beach pants

提示词：**beach pants, relaxed and casual, vibrant colors, walking along the sandy beach, palm trees swaying in the breeze, clear blue sky, a master anime illustrator, --ar 1:2 --niji 5**

兔子服装　bunny costume

提示词：bunny costume, cute and fluffy, anime style, big expressive eyes, in a whimsical forest setting, surrounded by flowers and butterflies, soft lighting, a master anime illustrator, ––ar 1:2 ––niji 5

圣葛罗莉安娜女学园校服　St. Gloriana's school uniform

提示词：St. Gloriana's school uniform, anime style,holding a bouquet of flowers, standing in a field of sunflowers, vibrant colors, sunny atmosphere, a master anime illustrator, ––ar 1:2 ––niji 5

鬼角　oni horns

提示词: oni horns, mysterious and alluring, flowing black gown, silver jewelry and intricate accessories, moonlit night setting, ethereal glow surrounding her, starry sky above, ancient ruins in the background, a master anime illustrator, ––ar 2:2 ––niji 5

机械光环　mechanical halo

提示词: mechanical halo, anime style, steampunk world, mysterious characters with goggles and top hats, a master anime illustrator, ––ar 2:2 ––niji 5

眼镜　glasses

提示词：girl wearing **glasses**, intelligent and studious, clear lenses, book in hand, a master anime illustrator, --ar 2:2 --niji 5

太阳镜　sunglasses

提示词：girl wearing **sunglasses**, stylish, intelligent and studious, dark lenses, a master anime illustrator, --ar 2:2 --niji 5

耳机　　headset

提示词：headset, anime style, futuristic setting, cyberpunk theme, neon lights, holographic projections, sleek and modern outfit, glowing headphones, intense expression, action pose, dark and gritty atmosphere, digital painting, a master anime illustrator, ––ar 2:2 ––niji 5

兽耳头罩　　animal hood

提示词：animal hood, cute and fluffy, pastel colors, oversized ears, sparkly eyes, kawaii expression, a master anime illustrator, ––ar 2:2 ––niji 5

鱼人耳　head fins

提示词: head fins, flowing and translucent, swirling patterns, anime style, composition, dynamic and energetic atmosphere, a master anime illustrator, --ar 2:2 --niji 5

头戴式显示器　head mounted display

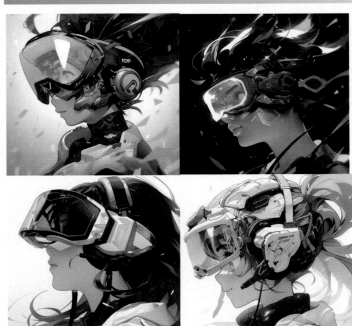

提示词: head mounted display, anime style, futuristic design, vibrant colors, sleek and streamlined, intense energy, a master anime illustrator, --ar 2:2 --niji 5

额前有宝石　forehead jewel

提示词：forehead jewel,sparkling and intricate design, colorful gemstones, delicate chains, ethereal and mystical atmosphere, flowing hair, a master anime illustrator, ––ar 2:2 ––niji 5

头上趴着猫　a cat sitting on the head

提示词：a cat sitting on the head, cute and fluffy, playful expression, anime style, big round eyes, whiskers, a master anime illustrator, ––ar 2:2 ––niji 5

半面罩　half mask

提示词：girl wearing a half mask, mysterious and alluring, eyes shining through the mask, flowing hair, soft and gentle expression, a master anime illustrator, --ar 2:2 --niji 5

头巾　bandana

提示词：bandana, red, tied around the head, worn by a biker, a master anime illustrator, --ar 2:2 --niji 5

头盔　helmet

提示词：helmet, man, rugged and worn-out appearance, battle scars and scratches, a master anime illustrator, --ar 2:2 --niji 5

皇冠　crown

提示词: girl wearing a crown, elegant and regal, flowing dress, surrounded by flowers, soft sunlight illuminating her face, a master anime illustrator, --ar 2:2 --niji 5

海盗帽　pirate hat

提示词：pirate hat, anime style, colorful feathers, skull and crossbones motif, tilted head to the side, wide brim, flowing ribbons, vibrant colors, exaggerated proportions, big eyes, dynamic pose, action-packed scene, in a manga-inspired art style, a master anime illustrator, --ar 2:2 --niji 5

魔女帽　witch hat

提示词: witch hat, anime style, cute and colorful, adorned with ribbons and bows, tilted head to the side, magical sparkles floating around it, a master anime illustrator, --ar 2:2 --niji 5

报童帽　newsboy hat

提示词：newsboy hat, anime style,vibrant colors, oversized brim, playful expression, strands of hair, dynamic pose, a master anime illustrator, --ar 2:2 --niji 5

草帽　straw hat

提示词：straw hat, girl, vintage style,wide brim, weathered and worn, floppy, straw texture, a master anime illustrator, --ar 2:2 --niji 5

边饰发带　trimmed hairband

提示词: trimmed hairband, anime style, pastel colors, lovely expression, flowing hair, a master anime illustrator, --ar 2:2 --niji 5

花环　flower necklace

提示词: flower necklace, girl with long flowing hair, vibrant colors, sunny outdoor setting, soft and dreamy atmosphere, a master anime illustrator, --ar 2:2 --niji 5

第6章

人物动作

人物动作是指角色在画面中的行为和姿势，它可以展示角色的性格、情绪、意图和能力等方面的特征。人物动作对故事情节的推进和角色关系的表现起着重要的作用。

站立　standing

提示词：**standing** anime character, sleek and futuristic outfit, cherry blossom trees in the background, soft and delicate color palette, peaceful atmosphere,a master anime illustrator, ––ar 1:2 ––niji 5

躺　lying

提示词：**lying**, anime style, relaxed and peaceful expression, colorful hair, oversized hoodie, surrounded by pillows and stuffed animals, soft and dreamy lighting, pastel color palette, floating cherry blossom petals, a master anime illustrator, ––ar 1:2 ––niji 5

跪　kneeling

提示词: **person kneeling, humble posture, hands clasped in prayer, serene expression, soft lighting, peaceful atmosphere, traditional clothing, wooden floor, minimalist composition, a master anime illustrator, --ar 1:2 --niji 5**

侧卧　lying on side

提示词: **anime style, girl lying on her side, peaceful and serene expression, flowing hair, soft pastel colors, dreamy atmosphere, surrounded by flowers and butterflies, delicate details, a master anime illustrator, --ar 1:2 --niji 5**

战斗姿态　fighting stance

提示词: **fighting stance**, anime, intense battle, dynamic poses, flowing hair, glowing eyes, powerful energy aura, dramatic lighting, vibrant colors, detailed background, a master anime illustrator, --ar 1:2 --niji 5

潜水　diving

提示词：a girl wearing a diving suit, diving, tropical fish swimming around, diving gear, goggles, underwater camera capturing the moment, serene and peaceful atmosphere, captured in a high-speed burst mode, a master anime illustrator, --ar 1:2 --niji 5

化妆　applying makeup

提示词：girl, applying makeup, delicate features, flawless skin, precise brush strokes, vibrant eyeshadow, perfectly winged eyeliner, rosy cheeks, glossy lips, a master anime illustrator, --ar 1:2 --niji 5

喷火　breathing fire

提示词：breathing fire, dragon, fierce, scales, majestic wings, sharp claws, glowing eyes, mythical creature, dark and stormy sky, epic battle scene, by J.R.R. Tolkien, --ar 1:2 --niji 5

扫帚骑行　broom riding

提示词：broom riding, anime style, school girl, witch, oversized hat, cheerful and energetic expression, flying through a colorful autumn forest, leaves swirling in the air, a master anime illustrator, ——ar 1:2 ——niji 5

追逐　chasing

提示词：chasing, anime girl and boy, running through a field of cherry blossom trees, wind blowing their hair and clothes, bright and colorful surroundings, sakura petals falling all around them, happy and carefree expressions, soft and dreamy atmosphere, ——ar 1:2 ——niji 5

拥抱　cuddling

提示词：cuddling, warm and cozy, soft blankets, gentle embrace, tender affection, peaceful atmosphere, dimmed lights, crackling fireplace, comfortable sofa, a master anime illustrator, --ar 1:2 --niji 5

跳舞　dancing

提示词：dancing, graceful movements, flowing dress, twirling and spinning, elegant posture, ballet shoes, soft and delicate lighting, dreamlike atmosphere, a master anime illustrator, --ar 1:2 --niji 5

祈祷　praying

提示词: standing **praying**, anime style, innocent expression, small hands pressed together, pigtails with ribbons, pastel tones, meadow background with flowers and butterflies, whimsical atmosphere, a master anime illustrator, −−ar 1:2 −−niji 5

走路　walking

提示词: **walking**, anime style, monochrome colors, rainy day, umbrella, melancholic atmosphere, long coat flowing in the wind, cityscape with neon lights reflecting on wet pavement, steam rising from the ground, mysterious atmosphere, a master anime illustrator, −−ar 1:2 −−niji 5

猫爪手势　cat pose

提示词：girl in cat pose, with an anime style, big round eyes, playful expression, long flowing hair, wearing a cute cat ear headband, surrounded by colorful flowers and butterflies, a master anime illustrator, ――ar 1:2 ――niji 5

攀爬　climbing

提示词：climbing girl, fearless adventurer, free-spirited expression, graceful movements, effortlessly scaling a towering cliff, a master anime illustrator, ――ar 1:2 ――niji 5

托脸颊　chin on hand

提示词：girl resting her chin on her hand, anime style, big bright eyes, rosy cheeks, soft pastel colors, flowing hair, dreamy background, sparkles and stars, cute and innocent expression, kawaii aesthetic, a master anime illustrator, ––ar 1:2 ––niji 5

抱着玩偶　holding doll

提示词：girl holding doll, with anime–style features, big round eyes, colorful hair, cute outfit, soft and huggable doll, in a whimsical fantasy world, surrounded by floating flowers and butterflies, in a pastel color palette, with a dreamy atmosphere, drawn with delicate lines and vibrant colors, a master anime illustrator, ––ar 1:2 ––niji 5

拿着面具　　holding mask

提示词：girl **holding mask**, vibrant colors, anime style, big expressive eyes, flowing hair, intricate details on the mask, dynamic pose, magical background with sparkling stars and moon, ethereal atmosphere, a master anime illustrator, --ar 1:2 --niji 5

手持法杖　　holding staff

提示词：girl **holding staff**, short bob haircut, fiery colors, fierce expression, surrounded by flames, volcanic landscape in the background, a master anime illustrator, --ar 1:2 --niji 5

唱跳　singing and dancing

提示词：singing and dancing, energetic dance group, colorful costumes, neon lights, disco ball, dynamic poses, glitter and confetti in the air, in a crowded concert hall, digital painting style, by Takashi Murakami, --ar 1:2 --niji 5

回眸　looking back

提示词：looking back, school uniform, book bag slung over shoulder, determined expression, bustling city street, autumn leaves falling, friends chatting in the background, sunset glow casting a warm light, gentle breeze blowing hair, in a slice-of-life style, --ar 1:2 --niji 5

高踢　high kick

提示词：**high kick**, dynamic and energetic, anime style, vibrant colors, exaggerated poses, flowing hair, intense facial expressions, background with motion lines and speed effects, a master anime illustrator, −−ar 1:2 −−niji 5

踮起脚　tiptoes

提示词：**tiptoes**, anime style, playful, joyful, dancing in the rain, splashing water droplets, carefree spirit, vibrant colors, laughter in the air, summer vibes, a master anime illustrator, −−ar 1:2 −−niji 5

第 7 章

场景氛围

场景氛围是指通过使用特定的色调和灯光效果来传达人物的情感。例如，冷色调的光线可以营造出冷静、阴郁或紧张的氛围，而暖色调的光线可以营造出温暖、宁静或浪漫的氛围。

霓虹暖光　neon warm lighting

提示词：neon warm lighting, vibrant colors, cherry blossom trees in full bloom, bustling street market with vendors selling street food and souvenirs,lanterns hanging overhead casting a warm glow on everything, ––ar 16:9 ––niji 5

温暖光辉　warm glow

提示词：warm glow, soft and golden sunset light, illuminating a field of blooming flowers, vibrant colors of petals, gentle breeze rustling through the grass, peaceful and joyful atmosphere, a master anime illustrator, ––ar 16:9 ––niji 5

梦幻雾气　dreamy haze

提示词：dreamy haze, anime style, soft pastel colors, ethereal atmosphere, flowing hair and clothing, glowing eyes, whimsical background with floating objects, magical powers or abilities, a master anime illustrator, ––ar 16:9 ––niji 5

魔法森林　enchanted forest

提示词：enchanted forest, anime style, glowing flowers that emit soft light, magical waterfalls that flow upwards, vibrant colors and surreal landscapes, a master anime illustrator, ––ar 16:9 ––niji 5

仙气缭绕　ethereal mist

提示词: ethereal mist, celestial beings, vibrant and surreal colors, glowing orbs of light, intricate and delicate details, a sense of wonder and awe, anime style, a master anime illustrator, ——ar 16:9 ——niji 5

柔和月光　soft moonlight

提示词: soft moonlight, Japanese-style, mystical forest with tall bamboo trees hidden shrine with torii gates leading to it, traditional kimono figure standing in front of the shrine with a sword in hand, ——ar 16:9 ——niji 5

浪漫烛光　romantic candlelight

提示词：romantic candlelight, soft and warm glow, flickering flames, cozy setting, vintage candlesticks, delicate candle holders, elegant tablecloth, fine china plates, crystal wine glasses, fresh flowers in a vase, a master anime illustrator, ––ar 16:9 ––niji 5

电光闪烁　electric flash

提示词：electric flash, intense brightness, blinding light, crackling sparks, electric arcs, glowing filaments, electrified air, a master anime illustrator, ––ar 16:9 ––niji 5

黄金时段光　golden hour light

提示词：golden hour light, anime style, soft and warm hues, magical and fantastical atmosphere, characters with supernatural abilities, a master anime illustrator, --ar 16:9 --niji 5

晨光　morning light

提示词：morning light, anime style, pastel colors, gentle and dreamy atmosphere, characters in school uniforms, cherry blossom petals floating in the air, soft breeze rustling through the trees, shy and innocent expressions, a master anime illustrator, --ar 16:9 --niji 5

沙滩　sandy beach

提示词：sandy beach, calm waves, clear blue water, palm trees swaying in the breeze, a gentle sea breeze, a peaceful and relaxing atmosphere, a master anime illustrator, --ar 16:9 --niji 5

幻想　fantasy

提示词：fantasy, glowing golden eyes, wearing the warrior's costume, standing on a over a tranquil pond in a serene forest, playful and mysterious atmosphere, a master anime illustrator, --ar 16:9 --niji 5

教室　classroom

提示词：classroom, vibrant colors, large windows, wooden desks and chairs, blackboard covered in equations and drawings,sunlight streaming in, students engaged in group discussions, lively and energetic atmosphere, a master anime illustrator, --ar 16:9 --niji 5

卧室　bedroom

提示词：bedroom, anime style, clean lines and simple furniture, a low platform bed with crisp white sheets and pillows, a wall-mounted bookshelf with neatly arranged books, a large window with blackout curtains, a master anime illustrator, --ar 16:9 --niji 5

森林　forest

提示词：**forest**, ancient and mysterious, foggy atmosphere, the moon cast a little light, anime style, a master anime illustrator, ––ar 16:9 ––niji 5

城市　city

提示词：**bustling city**, vibrant and colorful streets, towering skyscrapers, neon lights illuminating the night, a master anime illustrator, ––ar 16:9 ––niji 5

废弃城市建筑群　deserted city buildings

提示词：deserted city buildings, anime style, dilapidated structures, broken windows and doors, rusted metal beams, overgrown vines crawling up the walls, empty streets, a master anime illustrator, ——ar 16:9 ——niji 5

近未来都市　near future city

提示词：near future city, anime style, towering skyscrapers, futuristic technology, oversized billboards, holographic advertisements, detailed backgrounds with intricate architectural designs, dramatic lighting effects, a master anime illustrator, ——ar 16:9 ——niji 5

街景　street scenery

提示词：street scenery, sunset, golden hour lighting, bustling people crowded along the street, silhouettes of people walking, warm and soft color palette, a master anime illustrator, --ar 16:9 --niji 5

炼金室　Alchemy Laboratory

提示词：Alchemy Laboratory, anime style, whimsical and fantastical, filled with floating objects and levitating potions, vibrant colors and intricate patterns covering every surface, a master anime illustrator, --ar 16:9 --niji 5

咖啡厅　cafe

提示词：cafe, modern, minimalist design, sleek furniture, clean lines, large windows with city view, vibrant and energetic atmosphere, a master anime illustrator, ‒‒ar 16:9 ‒‒niji 5

居酒屋　Izakaya

提示词：Izakaya, modern twist, sleek and minimalist design, concrete walls and floors, industrial lighting fixtures, long communal tables with bar stools, open kitchen with chefs preparing dishes in front of customers, a master anime illustrator, ‒‒ar 16:9 ‒‒niji 5

实验室　laboratory

提示词：**laboratory, anime style, glowing test tubes, bubbling beakers, high–tech equipment, a young scientist in a white lab coat, surrounded by floating holographic screens displaying complex data ––ar 16:9 ––niji 5**

植物园　botanical garden

提示词：**botanical garden, lush greenery, vibrant flowers in full bloom, winding paths, tranquil ponds with floating lotus leaves, sunlight filtering through the trees, serene and peaceful atmosphere, a master anime illustrator, ––ar 16:9 ––niji 5**

游乐园　amusemen park

提示词：amusement park, vibrant colors, anime style characters, roller coasters, Ferris wheel, carousel, cotton candy, popcorn, laughter and excitement, joyful atmosphere, sunny day, a master anime illustrator, --ar 16:9 --niji 5

电车内　train interior

提示词：train interior, cozy and rustic atmosphere, warm wooden accents, soft ambient lighting, vintage-inspired details, peaceful and inviting vibe, a master anime illustrator, --ar 16:9 --niji 5

自动贩卖机　vending machine

提示词：vending machine, anime style, retro vibes, a quiet alleyway at night as the setting, composition focused on the vending machine as the central element, creating a sense of nostalgia and solitude, a master anime illustrator, ––ar 16:9 ––niji 5

鸟居　torii

提示词：torii, traditional Japanese gate, vibrant red color, wooden structure, located in a serene and peaceful garden, a master anime illustrator, ––ar 16:9 ––niji 5

情人节　valentine

提示词：**valentine, anime style, vibrant and bold colors, couples playing games and laughing, energetic and dynamic poses, a master anime illustrator, --ar 16:9 --niji 5**

万圣节　Halloween

提示词：**Halloween, anime style, cute witch girl, wearing a pointy hat, magical glowing pumpkins floating in the air, moonlit night with twinkling stars, a master anime illustrator, --ar 16:9 --niji 5**

圣诞节　Christmas

提示词：Christmas, cozy houses decorated with lights and ornaments, characters exchanging gifts and enjoying a feast, starry night sky, glowing lanterns lighting up the scene, a master anime illustrator, --ar 16:9 --niji 5

富士山　Mount Fuji

提示词：majestic Mount Fuji, traditional Japanese style, cherry blossom trees in full bloom, misty atmosphere, calm lake reflecting the mountain, wooden houses, --ar 16:9 --niji 5

东京塔　Tokyo Tower

提示词：Tokyo Tower, vibrant colors, anime style, tall and slender structure, glowing lights, bustling cityscape in the background, fluffy clouds in the sky, a master anime illustrator, ––ar 16:9 ––niji 5

书店　bookstore

提示词：bookstore, cozy, bookshelves filled with books, soft lighting, vintage furniture, warm colors, a quiet and peaceful atmosphere, a master anime illustrator, ––ar 16:9 ––niji 5

体育场   stadium

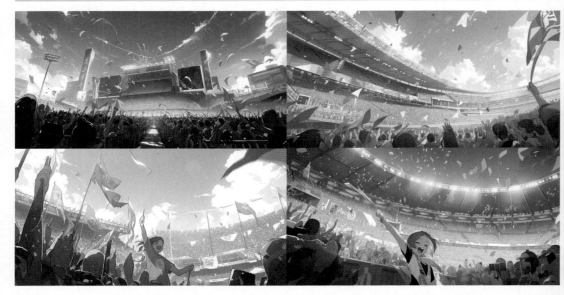

提示词：stadium, packed with cheering fans, vibrant energy, colorful banners and flags, intense competition, roaring crowd, floodlights illuminating the field, dynamic action shots, a master anime illustrator, ––ar 16:9 ––niji 5

游泳馆   swimming pool

提示词：swimming pool, Olympic-sized, professional swimming pool, a master anime illustrator, ––ar 16:9 ––niji 5